Maths Problem Solving

Year 6

Catherine Yemm

Other books in the same series:

Maths Problem Solving – Year 1   ISBN 10... 1 903853 745
                                 ISBN 13... 978-1-903853-74-0
Maths Problem Solving – Year 2   ISBN 10... 1 903853 753
                                 ISBN 13... 978-1-90385375-7
Maths Problem Solving – Year 3   ISBN 10... 1 903853 761
                                 ISBN 13... 978-1-903853-76-4
Maths Problem Solving – Year 4   ISBN 10... 1 903853 77X
                                 ISBN 13... 978-1-903853-77-1
Maths Problem Solving – Year 5   ISBN 10... 1 903853 788
                                 ISBN 13... 978-1-903853-78-8

Published by Brilliant Publications
1 Church View, Sparrow Hall Farm, Edlesborough,
Dunstable, Bedfordshire LU6 2ES

Sales and stock enquiries	Tel:	01202 712910 / 0845 1309200
	Fax:	0845 1309300
Sales and payments	E-mail:	brilliant@bebc.co.uk
	website:	www.brilliantpublications.co.uk
General enquiries:	Tel:	01525 229720

The name Brilliant Publications and its logo are registered trade marks.

Written by Catherine Yemm
Cover and illustrations by Frank Endersby

ISBN 1 903853 796
ISBN 13... 978-1-903853-79-5
© Catherine Yemm 2005
First published in 2005
Printed in the UK by Lightning Source
10   9   8   7   6   5   4   3   2   1

Contents

Introduction

Maths Problem Solving Year 6 is the last book in a series of six resource books written for teachers to use during the Numeracy lesson. It specifically covers the objectives from the Numeracy framework that are collated under the heading 'Solving problems'. Each book is specific for a particular year group and contains clear photocopiable resources which can be photocopied onto acetate to be viewed by the whole class or photocopied onto paper to be used by individuals.

Problem solving plays a very important part in the Numeracy curriculum and one of the reasons Numeracy is such an important subject is because the skills the children learn enable them to solve problems in other aspects of their lives. It is not enough to be able to count, recognize numbers and calculate; children need to be able to use problem solving skills alongside mathematical knowledge to help them succeed in a variety of 'real life' situations. Many of the problem solving skills and strategies that are needed do not come naturally so they have to be taught.

Problem solving is not an area which should be taught exclusively on its own but one area which should be taught alongside other mathematical areas such as numbers and shape, space and measures. Children will benefit from being given opportunities to solve problems in other areas of the curriculum and away from the classroom as well as in their Numeracy lessons.

When teaching children how to solve problems the Numeracy strategy refers to a number of points that need to be considered:

- The length of the problems should be varied depending on the age group. Children will benefit from being given short, medium-length and more extended problems.
- The problems on one page or in one lesson should be mixed so that the children do not just assume they are all 'multiplication' problems, for example, and simply multiply the numbers they see to find each answer.
- The problems need to be varied in their complexity: there should be some one-step and some two-step problems, and the vocabulary used in each problem should differ.
- Depending on the age of the children the problems can be given orally or in writing. When given written problems to solve, some children may need help to read the words, although this does not necessarily mean that they will need help to find the answer to the question.
- The context of the problem should be meaningful and relevant to the children. It should attempt to motivate them into finding the answer and be significant to the time. For example, you could introduce euros into the questions.

This Year 6 resource is organized into three chapters: 'Making decisions', 'Reasoning about numbers or shapes' and 'Problems involving "real life", money or measures'. Each chapter contains six lessons, one to be used each half term.

Making decisions

The objective outlined under the 'Making decisions' heading of the National Numeracy Strategy for Year 6 children is: Choose and use appropriate number operations to solve problems, and find appropriate ways for calculating, for example mental, mental with jottings, written methods, calculator.

In this chapter the emphasis is on choosing and then using the correct operation to solve a given problem. In Year 6 the children are developing their adding, subtracting, multiplying and dividing skills so they need to understand that different problems will need different methods to solve them. They should be encouraged to make and justify decisions by choosing the appropriate operations to solve word problems, deciding whether calculations can be done mentally or by using a pencil and paper or by explaining and recording how the problem was solved. The children should be provided with an opportunity to tackle mixed problems so that they learn to think openly and make a decision depending on the vocabulary used in the question itself. If children are not taught these decisive skills then it is common for them to assume that, to find the answer to a question with two numbers, you just add or multiply the numbers. The questions set out in this chapter are mixed and the children could be required to use any of the four operations. The questions the children will answer are designed to enable them to practise solving problems in a variety of relevant contexts.

When the children are answering the questions encourage them to use mental strategies to make notes and use more formal written methods and give them an opportunity to use a calculator.

This aspect of problem solving is closely correlated to objectives 72–73, 'Checking results of calculations'. After choosing and using the correct operation the children should be encouraged, where appropriate, to use a method to check their results, which could be: using the inverse operation; adding in the reverse order; using an equivalent calculation; approximating or using knowledge of sums, differences and products of odd or even numbers; using tests of divisibility.

Reasoning about numbers or shapes

The objectives outlined under the 'Reasoning about numbers or shapes' heading of the National Numeracy Strategy for Year 6 children are as follows:
- Explain methods and reasoning orally and, where appropriate, in writing
- Solve mathematical problems or puzzles; recognize and explain patterns and relationships, generalize and predict. Suggest extensions by asking 'What if … ?'

- Make and investigate a general statement about familiar numbers or shapes by finding examples that satisfy it. Develop from explaining a generalized relationship in words to expressing it in a formula using letters as symbols.

The activities in this chapter are a mixture of problems, puzzles and statements. Lessons 1, 3 and 5 are related to shape while lessons 2, 4 and 6 are related to number. When given a statement such as: 'If you multiply a number by a half it makes the number twice as small', the children should be encouraged to provide examples to prove the statement, eg $32 \times 0.5 = 16$; $72 \times 0.5 = 36$. Others may be obvious questions that just need an answer. The teacher should try to spend time talking to the pupils while they are working to allow them to explain their methods and reasoning orally and to provide an opportunity to ask questions such as 'What if... ?' The plenary session at the end of the lesson also provides an opportunity to do this.

Problems involving 'real life', money or measures

The objectives outlined under the 'Problems involving 'real life', money and measures' heading of the National Numeracy Strategy for Year 6 children are as follows:
- Identify and use appropriate operations (including combinations of operations) to solve word problems involving numbers and quantities based on 'real life', money or measures (including time), using one or more steps to convert pounds to foreign currency, or *vice-versa*, and calculating percentages such as VAT.
- Explain methods and reasoning. The activities in this chapter are typically 'word problems'. The contexts are designed to be realistic and relevant for children of Year 6 age. The questions involve the operations of adding, subtracting, multiplication and division, and deal with money, measurements including time and everyday situations.

The teacher should try to spend time talking to the pupils while they are working to allow them to explain their methods and reasoning orally. The plenary session at the end of the lesson also provides an opportunity to do this.

This aspect of problem solving is closely correlated to objectives 72–73, 'Checking results of calculations'. After choosing and using the correct operation the children should be encouraged, where appropriate, to use a method to check their results, which could be: using the inverse operation; adding in the reverse order; using an equivalent calculation; approximating or using knowledge of sums, differences and products of odd or even numbers; using tests of divisibility.

The lesson

Mental starter

In line with the Numeracy strategy the teacher should start the lesson with a 5–10 minute mental starter. This can be practice of a specific mental skill from the list specified for that particular half term or ideally an objective linked to the problems the children will be solving in the main part of the lesson. For example, if the problems require the children to add and subtract then it would be useful to spend the first 10 minutes of the lesson reinforcing addition and subtraction bonds and the vocabulary involved.

The main teaching activity and pupil activity

This book aims to provide all the worksheets that the teacher will need to cover this part of the lesson successfully. The first page has examples of the types of problems that need to be solved. It is designed to be photocopied onto acetate to show the whole class using an overhead projector. The teacher will use the empty space to work through the examples with the class before introducing them to questions they can try by themselves. The teacher should demonstrate solving the problem using skills that are relevant to the abilities of the children in the class, such as using drawings, counters or number lines.

Once the children have seen a number of examples they will be ready to try some problem solving questions for themselves. Within each lesson there is a choice of three differentiated worksheets. The wording for each question on these worksheets is the same, but the level of mathematical complexity varies. This ensures that the questions are differentiated only according to the mathematical ability of the children. It will also ensure that when going through examples during the plenary session all children can be involved at the same time. For example, for a question involving the addition of three numbers, the children may have had to add three different numbers, but when the teacher talks through the question the fact that to solve the problem the children need to add will be the important point being reinforced. If children are all completing totally different types of questions then when the teacher talks through a question in the plenary session some groups of children may have to sit idle as they did not have that question on their sheet.

If the teacher feels that some pupils would benefit from having easier or more difficult questions then they could change the numbers on the worksheets to something more suitable.

The plenary

One of the important parts of solving problems is discussing how problems were solved and the plenary lends itself well to this. After the children have completed the problems the plenary can be used to:

- discuss the vocabulary used in the question
- discuss how the problem can be approached
- break down a problem into smaller steps
- list the operations or calculations used to solve the problem
- discuss if the problem could be solved in more than one way
- discuss how the answers to the problems can be checked
- divulge the answers to a number of the questions.

Extension

Any children who complete their task relatively easily may need to be extended further. As well as being given the more challenging questions they could be asked to make up a question of their own, which should involve the same operations.

Resources

For some questions it will be useful to make a number of resources available to the children such as:

- Number lines to 100
- A selection of 2D and 3D shapes
- All coins
- Analogue clocks with moveable hands, digital clocks.

Answers

We have supplied answers to questions where possible, but there are some questions which have multiple answers or require class discussion. Some questions require the children to show their understanding by making up a story involving the figures mentioned, and some others are statements which require the children to give an example which supports the fact.

Photocopiable answer sheet

Photocopy onto acetate sheet and project on wall or screen.

I will need to ....

To help me I will use ....

My answer is ....

I will need to ....

To help me I will use ....

My answer is ....

I will need to ....

To help me I will use ....

My answer is ....

Making decisions

Whole class activity

In June the sun shines for 13 hours and 23 minutes a day. In the winter it shines for 15,300 seconds less. For how many hours and minutes does it shine?

In one year 2 430 children swim at the local swimming pool. If they pay £1.25 each how much does the leisure centre make from the children's swimming?

What operation does the * represent?

3740 * 55 = 68

How can you check?

This page may be photocopied by the purchasing institution only.

© Catherine Yemm

1. Make up a number story to reflect this calculation:

   57.9 + 28.6 = 86.5

   •

2. The local greengrocer sells 12 cherries in a bag. If he sells 13 bags in one day how many cherries will he have sold?

   •

3. I think of a number. I add 2.3 and multiply it by 3. My answer is 156.9. What was my number?

   •

4. Danielle and Natasha each do a paper round. Danielle has saved £31.56 from her round and Natasha has saved £82.17 from her round. How much money do they have between them?

   •

5. Juliet's bottle of water holds 250ml. Craig's bottle holds 2.5 times as much. How much does it hold?

   •

6. What is the difference between 1.26 and 2.15?

   •

7. Chay has £10 to spend at the funfair. How much change will he have if he has 5 rides on the helter-skelter and it costs £1.10 a go?

   •

8. A local artist has donated 264 crayons to the school. If there are 12 classes in the school how many crayons can each class have? Will there be any left over?

   •

9. Miriam puts her cherry cake in the oven at 16:45. It needs to cook for 2 hours and 10 minutes. What time should she get it out?

   •

10. A school has 110 dictionaries. There are 3 infant classes and 4 junior classes. The junior classes need twice as many dictionaries as the infant classes. How many dictionaries can each infant and junior class have?

Make up a number story to reflect this calculation:

1. 142.9 + 28.65 = 171.55

• •

2. The local greengrocer sells 24 cherries in a bag. If he sells 13 bags in one day how many cherries will he have sold?

• •

3. I think of a number. I add 7.3 and multiply it by 6. My answer is 313.8. What was my number?

• •

4. Danielle and Natasha each do a paper round. Danielle has saved £131.56 from her round and Natasha has saved £182.17 from her round. How much money do they have between them?

• •

5. Juliet's bottle of water holds 250ml. Craig's bottle holds 3.25 times as much. How much does it hold?

• •

6. What is the difference between 2.26 and 3.92?

• •

7. Chay has £20 to spend at the funfair. How much change will he have if he has 5 rides on the helter-skelter and it costs £1.65 a go?

• •

8. A local artist has donated 455 crayons to the school. If there are 12 classes in the school how many crayons can each class have? Will there be any left over?

• •

9. Miriam puts her cherry cake in the oven at 16:43. It needs to cook for 2 hours and 8 minutes. What time should she get it out?

• •

10. A school has 198 dictionaries. There are 3 infant classes and 4 junior classes. The junior classes need twice as many dictionaries as the infant classes. How many dictionaries can each infant and junior class have?

1. Make up a number story to reflect the calculation:

    242.9 + 128.65 = 371.55

    ●

2. The local greengrocer sells 24 cherries in a bag. If he sells 18 bags in one day how many cherries will he have sold?

    ●

3. I think of a number. I add 7.3 and multiply it by 9. My answer is 515.7. What was my number?

    ●

4. Danielle and Natasha each do a paper round. Danielle has saved £241.56 from her round and Natasha has saved £282.17 from her round. How much money do they have between them?

    ●

5. Juliet's bottle of water holds 250ml. Craig's bottle holds 5.25 times as much. How much does it hold?

    ●

6. What is the difference between 2.16 and 6.92?

    ●

7. Chay has £20 to spend at the funfair. How much change will he have if he has 6 rides on the helter-skelter and it costs £1.93 a go?

    ●

8. A local artist has donated 874 crayons to the school. If there are 12 classes in the school how many crayons can each class have? Will there be any left over?

    ●

9. Miriam puts her cherry cake in the oven at 16:43. It needs to cook for 4 hours and 28 minutes. What time should she get it out?

    ●

10. A school has 308 dictionaries. There are 3 infant classes and 4 junior classes. The junior classes need twice as many dictionaries as the infant classes. How many dictionaries can each infant and junior class have?

Making decisions

Whole class activity

I think of a number and I subtract 55 and divide it by 2.5. My answer is 998. What was my number?

It costs £6.50 to visit the local art gallery. How much would it cost to take a class of 38 children?

If I increase 27.4 by 5.56 what number will I get?

© Catherine Yemm

1. In one week the amount of rain that falls from Monday to Sunday doubles every day. If it rains 1.1mm on Monday how much rain falls during the whole week?

• •

2. What operation does the * represent?
92 * 36 = 3312
How can you check?

• •

3. How many times 35 is 245?

• •

4. Make up a number story to reflect this calculation:

39.4 – 18.25 = 21.15

• •

5. Jamal's play pool usually holds 943 litres of water but there has been a leak and 232 litres have leaked out. How much water is left in the pool?

• •

6. Max has a 1200 piece jigsaw. He drops it on the floor and loses 350 pieces. How many does he have left?

• •

7. I think of a number and add 32 then I multiply it by 2.5. My answer is 130. What was my number?

• •

8. A school usually has 275 pupils on the register. 14 new families have moved to the area over the summer and they all have three children. How many children will be at the school in September?

• •

9. The train on platform 2 is supposed to arrive at 08:20 but it is running 35 minutes late. What time will it arrive?

• •

10. The local park is 137 metres long but it is shortened by 26.5 metres. How long is it now?

Lesson 2b

1. In one week the amount of rain that falls from Monday to Sunday doubles every day. If it rains 2.1mm on Monday, how much rain falls during the whole week?

• •

2. What operation does the * represent?

492 * 36 = 17 712

How can you check?

• •

3. How many times 35 is 385?

• •

4. Make up a number story to reflect this calculation:

59.4 − 18.25 = 41.15

• •

5. Jamal's play pool usually holds 2543 litres of water but there has been a leak and 432 litres have leaked out. How much water is left in the pool?

• •

6. Max has a 2200 piece jigsaw. He drops it on the floor and loses 476 pieces. How many does he have left?

• •

7. I think of a number and add 73 then I multiply it by 3.5. My answer is 325.5. What was my number?

• •

8. A school usually has 879 pupils on the register. 27 new families have moved to the area over the summer and they all have three children. How many children will be at the school in September?

• •

9. The train on platform 2 is supposed to arrive at 08:37 but it is running 48 minutes late. What time will it arrive?

• •

10. The local park is 237 metres long but it is shortened by 56.5 metres. How long is it now?

1. In one week the amount of rain that falls from Monday to Sunday doubles every day. If it rains 4.1mm on Monday, how much rain falls during the whole week?

2. What operation does the * represent?
692 * 32 = 22 144
How can you check?

3. How many times 35 is 875?

4. Make up a number story to reflect this calculation:

159.4 – 118.25 = 41.15

5. Jamal's play pool usually holds 22 543 litres of water but there has been a leak and 1432 litres have leaked out. How much water is left in the pool?

6. Max has a 4700 piece jigsaw. He drops it on the floor and loses 1276 pieces. How many does he have left?

7. I think of a number and add 93 then I multiply it by 5.5. My answer is 786.5 What was my number?

8. A school usually has 1879 pupils on the register. 57 new families have moved to the area over the summer and they all have three children. How many children will be at the school in September?

9. The train on platform 2 is supposed to arrive at 06:37 but it is running 2 hours and 48 minutes late. What time will it arrive?

10. The local park is 1237 metres long but it is shortened by 156.5 metres. How long is it now?

Making decisions

Whole class activity

How many more grams of apples does Efia need to add to 13.4kg to make 15kg?

Sara's bedroom has an area of 45m². Her older sister's bedroom has an area 2.5 times as big. What is the area of her sister's bedroom?

Geraldine's gran lives in Australia. On Gran's birthday Geraldine rings her in the morning and evening. The first phone call lasts 1247 seconds; the second lasts 4738 seconds. How many minutes and seconds does she spend talking to her gran that day?

1. Regan's piano lessons start at 17:25 and last for 25 minutes. She is having a double lesson as she has her exam soon. What time will she finish the lesson?

2. The village fête is providing tables at which people can have a drink. They are expecting 105 people and they have 15 tables. How many chairs should they put around each table?

3. What operation does the * represent?
251* 487 = 738
How can you check?

4. I think of a number. I add 341 and subtract 295. My answer is 300. What was my number?

5. In a writing competition 8 people shared the prize and won £52.75 each. How much was the total prize money?

6. How many more than 1892 is 2251?

7. Alex's grandma likes knitting. She can usually knit 476 stitches in an hour. At the moment she has a sore arm so she can only knit a quarter of the stitches in an hour. How many stitches less is she doing per hour?

8. Make up a number story to reflect the calculation below.
37.15 ÷ 5 = 7.43

9. The apple tree in Tharman's garden is 15.7 metres tall. The cherry tree is 250cm shorter. How tall is the cherry tree?

10. Nicola has read both her reading books this week. The first book had 152 pages and the second had half that number. How many pages has she read this week?

Making decisions

1. Regan's piano lessons start at 17:25 and last for 55 minutes. She is having a double lesson as she has her exam soon. What time will she finish the lesson?

• •

2. The village fête is providing tables at which people can have a drink. They are expecting 135 people and they have 15 tables. How many chairs should they put around each table?

• •

3. What operation does the * represent?
651 * 987 = 1638
How can you check?

• •

4. I think of a number. I add 441 and subtract 395. My answer is 400. What was my number?

• •

5. In a writing competition 12 people shared the prize and won £152.75 each. How much was the total prize money?

• •

6. How many more than 1892 is 3251?

• •

7. Alex's grandma likes knitting. She can usually knit 3476 stitches in an hour. At the moment she has a sore arm so she can only knit a quarter of the stitches in an hour. How many stitches less is she doing per hour?

• •

8. Make up a number story to reflect the calculation below.
401.22 ÷ 54 = 7.43

• •

9. The apple tree in Tharman's garden is 25.7 metres tall. The cherry tree is 450cm shorter. How tall is the cherry tree?

• •

10. Nicola has read both her reading books this week. The first book had 452 pages and the second had half that number. How many pages has she read this week?

1. Regan's piano lessons start at 17:25 and last for 65 minutes. She is having a double lesson as she has her exam soon. What time will she finish the lesson?

• •

2. The village fête is providing tables at which people can have a drink. They are expecting 195 people and they have 15 tables. How many chairs should they put around each table?

• •

3. What operation does the * represent?
1651 * 1987 = 3638
How can you check?

• •

4. I think of a number. I add 641 and subtract 495. My answer is 500. What was my number?

• •

5. In a writing competition 12 people shared the prize and won £252.75 each. How much was the total prize money?

• •

6. How many more than 6251 is 8892?

• •

7. Alex's grandma likes knitting. She can usually knit 8476 stitches in an hour. At the moment she has a sore arm so she can only knit a quarter of the stitches in an hour. How many stitches less is she doing per hour?

• •

8. Make up a number story to reflect the calculation below.
549.82 ÷ 74 = 7.43

• •

9. The apple tree in Tharman's garden is 55.7 metres tall. The cherry tree is 1450cm shorter. How tall is the cherry tree?

• •

10. Nicola has read both her reading books this week. The first book had 752 pages and the second had half that number. How many pages has she read this week?

Making decisions

Whole class activity

How many seconds are there in 3 hours and 17 minutes?

The carpet in Rupert's bedroom is 2.75 metres long. In his brother's bedroom, the carpet is 112cm shorter. How long is his brother's carpet?

The school cook buys 350 bananas for the week. 20% of the bananas are bruised and need to be thrown away. How many bananas are left?

1. Fred's mum buys 16 tins of fruit at the supermarket. 25% of the tins are past their sell by date and need to be thrown away. How many tins of fruit are left?

2. Devi has won £30 of book tokens in a competition. If he buys a new dictionary for £12.50 and an encyclopaedia for £15.95, how much money will he have left?

3. How many altogether are 858, 166 and 26.5?

4. It costs £1.25 per person to visit the local farm and they have an average of 220 visitors a month. How much money do they make in a year?

5. The members of the school relay team completed their race in the following times: swimmer 1, 34 seconds; swimmer 2, 25 seconds; swimmer 3, 83 seconds; and swimmer 4, 96 seconds. What was their overall time in minutes and seconds?

6. I think of a number and add 52.4 and then multiply it by 5. My answer is 674. What was my number?

7. What operation does the * represent?
925 * 435 = 490
How can you check?

8. A pencil is made from 235mm of lead. How many pencils can be made from 2 metres of lead?

9. The perimeter fence around the local primary school is 175 metres. The perimeter fence around the local high school is double that length. How long is it?

10. Make up a number story to reflect the calculation below.

7.13 x 18 = 128.34

Lesson 4b

1. Fred's mum buys 32 tins of fruit at the supermarket. 25% of the tins are past their sell by date and need to be thrown away. How many tins of fruit are left?

2. Devi has won £50 of book tokens in a competition. If he buys a new dictionary for £12.50 and an encyclopaedia for £27.95, how much money will he have left?

3. How many altogether are 1058, 266 and 26.5?

4. It costs £1.25 per person to visit the local farm and they have an average of 560 visitors a month. How much money do they make in a year?

5. The members of the school relay team completed their race in the following times: swimmer 1, 134 seconds; swimmer 2, 205 seconds; swimmer 3, 323 seconds; and swimmer 4, 196 seconds. What was their overall time in minutes and seconds?

6. I think of a number and add 32.4 and then multiply it by 5. My answer is 574. What was my number?

7. What operation does the * represent?
1925 * 1435 = 490
How can you check?

8. A pencil is made from 135mm of lead. How many pencils can be made from 2 metres of lead?

9. The perimeter fence around the local primary school is 275 metres. The perimeter fence around the local high school is double that length. How long is it?

10. Make up a number story to reflect the calculation below.

7.13 x 28 = 199.64

1. Fred's mum buys 64 tins of fruit at the supermarket. 25% of the tins are past their sell by date and need to be thrown away. How many tins of fruit are left?

• •

2. Devi has won £70 of book tokens in a competition. If he buys a new dictionary for £22.50 and an encyclopaedia for £27.95, how much money will he have left?

• •

3. How many altogether are 1258, 466 and 86.5?

• •

4. It costs £2.25 per person to visit the local farm and they have an average of 560 visitors a month. How much money do they make in a year?

• •

5. The members of the school relay team completed their race in the following times: swimmer 1, 184 seconds; swimmer 2, 275 seconds; swimmer 3, 423 seconds; and swimmer 4, 296 seconds. What was their overall time in minutes and seconds?

• •

6. I think of a number and add 82.4 and then multiply it by 5. My answer is 874. What was my number?

• •

7. What operation does the * represent?
2925 * 1435 = 1490
How can you check?

• •

8. A pencil is made from 135mm of lead. How many pencils can be made from 5 metres of lead?

• •

9. The perimeter fence around the local primary school is 475 metres. The perimeter fence around the edge of the local high school is double that length. How long is it?

• •

10. Make up a number story to reflect the calculation below.

17.13 x 28 = 479.64

Making decisions

In the leisure centre car park there are 32 rows of cars. Each row has 12 cars. How many cars are in the car park?

Tara, Tilak and Shauna collect stamps. Tara has 1423, Tilak has 312 less than Tara, and Shauna has 231 less than Tilak. How many stamps do they each have and how many do they have between them?

In the school hall there are 5 black chairs and 4 red chairs in every row. If the caretaker has put out 45 black chairs how many red chairs are there?

1. Four 44 seater buses turn up at school to take the children on a trip but in each bus a quarter of the seats are damaged. How many children can ride on the buses?

• •

2. The taps on Sheila's bath pour water at a rate of 2.5 litres per 30 seconds. How long will it take her to fill her bath with 100 litres of water?

• •

3. I think of a number I subtract 32 and divide it by 10. My answer is 1.2. What was my number?

• •

4. Make up a number story to reflect this calculation:

845 – 231 = 614

• •

5. How many more pence does Sophie need to add to £46.35 to make £50?

• •

6. The local leisure centre has 1674 tickets to give away to see a local band play in concert. This is twice the number of tickets as last year. How many people came to see the band play last year?

• •

7. What operation does the * represent?
403 * 26 = 15.50
How can you check?

• •

8. Jessica's tent weighs 86kg. Sarah's tent is a quarter of the weight. How heavy is Sarah's tent?

• •

9. Start at 784 and add a number to make 1205. What is the number?

• •

10. On the beach are 8 rows of deckchairs. Each row has 28 chairs. How many deckchairs are on the beach?

Making decisions

1. Four 84 seater buses turn up at school to take the children on a trip but in each bus a quarter of the seats are damaged. How many children can ride on the buses?

2. The taps on Sheila's bath pour water at a rate of 2.5 litres per 30 seconds. How long will it take her to fill her bath with 300 litres of water?

3. I think of a number. I subtract 32 and divide it by 20. My answer is 1.2. What was my number?

4. Make up a number story to reflect this calculation:

   9845 − 6231 = 3614

5. How many more pence does Sophie need to add to £76.35 to make £90?

6. The local leisure centre has 5674 tickets to give away to see a local band play in concert. This is twice the number of tickets as last year. How many people came to see the band play last year?

7. What operation does the * represent?
   854 * 56 = 15.25
   How can you check?

8. Jessica's tent weighs 134kg. Sarah's tent is a quarter of the weight. How heavy is Sarah's tent?

9. Start at 1284 and add a number to make 2084. What is the number?

10. On the beach are 16 rows of deckchairs. Each row has 28 chairs. How many deckchairs are on the beach?

Lesson
5c

1. Four 124 seater buses turn up at school to take the children on a trip but in each bus a quarter of the seats are damaged. How many children can ride on the buses?

2. The taps on Sheila's bath pour water at a rate of 2.5 litres per 30 seconds. How long will it take her to fill her bath with 600 litres of water?

3. I think of a number. I subtract 32 and divide it by 40. My answer is 1.2. What was my number?

4. Make up a number story to reflect this calculation:

   19 845 – 16 231 = 3614

5. How many more pence does Sophie need to add to £176.35 to make £190?

6. The local leisure centre has 15 674 tickets to give away to see a local band play in concert. This is twice the number of tickets as last year. How many people came to see the band play last year?

7. What operation does the * represent?
   976 * 64 = 15.25
   How can you check?

8. Jessica's tent weighs 234kg. Sarah's tent is a quarter of the weight. How heavy is Sarah's tent?

9. Start at 10 284 and add a number to make 21 084. What is the number?

10. On the beach are 26 rows of deckchairs. Each row has 28 chairs. How many deckchairs are on the beach?

Making decisions

Whole class activity

An apple pie needs to be cooked for 22 minutes for every kilogram it weighs. How long should it be cooked for if it weighs 3.5kg?

A full bottle of lemonade holds 4.5 litres. A cup holds 0.25 litres. How many cups of lemonade fit in the bottle?

In the local art gallery there are 29 display stands and on each there are 45 paintings. How many paintings are in the art gallery?

1. On holiday in France Martin bought a new surf board for 45.88 euros. Sam bought one in London for £42.25. If there are 1.48 euros to £1 whose board was cheaper and by how much?

    •

2. If 1 yard is 36 inches how many inches is 9 yards?

    •

3. What operation does the * represent?
    168 * 31 = 5208
    How can you check?

    •

4. Indra wants to buy a new pair of trainers that cost £84. Her dad has said that she can have a pair that are a third of the price. How much are the trainers that her dad wants to buy?

    •

5. How many centimetres of rope does Claudia need to cut off her 24.7 metre rope to make it a 17.4 metre rope?

    •

6. Which three numbers could have a total of 4.5?

    •

7. Philip's new dictionary has 874 definitions in it. Amy's only has 237. How many more definitions does Philip's dictionary have?

    •

8. Make up a number story to reflect the calculation below.

    243 x 16 = 3888

    •

9. I think of a number. I multiply it by 3 and subtract 45. My answer is 180. What was my number?

    •

10. There are 561 people who live in the local village. Two-thirds of these are female. How many are male?

Lesson 6b

1. On holiday in France Martin bought a new surf board for 115.44 euros. Sam bought one in London for £88.25. If there are 1.48 euros to £1 whose board was cheaper and by how much?

2. If 1 yard is 36 inches how many inches is 15 yards?

3. What operation does the * represent?
268 * 31 = 8308
How can you check?

4. Indra wants to buy a new pair of trainers that cost £102. Her dad has said that she can have a pair that are a third of the price. How much are the trainers that her dad wants to buy?

5. How many centimetres of rope does Claudia need to cut off her 34.7 metre rope to make it a 17.4 metre rope?

6. Which three numbers could have a total of 8.95?

7. Philip's new dictionary has 2874 definitions in it. Amy's only has 1237. How many more definitions does Philip's dictionary have?

8. Make up a number story to reflect the calculation below.

243 x 56 = 13 608

9. I think of a number. I multiply it by 3 and subtract 50. My answer is 280. What was my number?

10. There are 1134 people who live in the local village. Two-thirds of these are female. How many are male?

1. On holiday in France Martin bought a new surf board for 222 euros. Sam bought one in London for £188.25. If there are 1.48 euros to £1 whose board was cheaper and by how much?

• •

2. If 1 yard is 36 inches how many inches is 21 yards?

• •

3. What operation does the * represent?
368 * 31 = 11408

How can you check?

• •

4. Indra wants to buy a new pair of trainers that cost £144. Her dad has said that she can have a pair that are a third of the price. How much are the trainers that her dad wants to buy?

• •

5. How many centimetres of rope does Claudia need to cut off her 44.7 metre rope to make it a 17.4 metre rope?

• •

6. Which three numbers could have a total of 15.72?

• •

7. Philip's new dictionary has 3874 definitions in it. Amy's only has 1237. How many more definitions does Philip's dictionary have?

• •

8. Make up a number story to reflect the calculation below.

243 x 56 = 13 608

• •

9. I think of a number. I multiply it by 3 and subtract 50. My answer is 355. What was my number?

• •

10. There are 1566 people who live in the local village. Two-thirds of these are female. How many are male?

Reasoning about numbers or shapes

Whole class activity

What could the time be if the hands on an analogue clock show an angle of 150⁰?

How many of these triangles do you need to make a triangle that is 9 times the size?

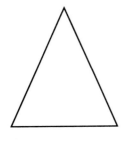

These 4 lines make **a square**. How many squares can you make with 20 lines?

1. Draw and name a 3D shape that has more than 8 faces.

· ·

2. How many right angles are there in this picture?

· ·

3. Cut a trapezium in half. What shapes do you get?

· ·

4. What could the sizes of the three angles in a triangle be?

· ·

5. How many small cubes would you need to make a cube that is 16 cubes wide?

· ·

6. Draw an irregular nonagon.

· ·

7. What do the inside angles of a regular pentagon measure?

· ·

8. Does the net of a heptagonal prism have more vertices than the net of a pentagonal prism?

· ·

9. How many lines of symmetry do the letters in the word 'SYMMETRY' have altogether?

· ·

10. Which shapes would you need to make a 3D model of a hexagonal prism?

Lesson 1b

1. Draw and name a 3D shape that has more than 9 faces.

• •

2. How many right angles are there in this picture?

3. Cut a trapezium in three. What shapes do you get?

• •

4. What could the sizes of the five angles in a pentagon be?

• •

5. How many small cubes would you need to make a cube that is 20 cubes wide?

• •

6. Draw an irregular decagon.

• •

7. What do the inside angles of a regular hexagon measure?

• •

8. Does the net of a hexagonal prism have more vertices than the net of a heptagonal prism?

• •

9. How many lines of symmetry do the letters in the word 'MAGNIFICENT' have altogether?

• •

10. Which shapes would you need to make a 3D model of a heptagonal prism?

1. Draw and name a 3D shape that has more than 10 faces.

2. How many right angles are there in this picture?

3. Cut a trapezium in four. What shapes do you get?

4. What could the sizes of the eight angles in an octagon be?

5. How many small cubes would you need to make a cube that is 25 cubes wide?

6. Draw an irregular dodecagon.

7. What do the inside angles of a regular heptagon measure?

8. Does the net of a heptagonal prism have more vertices than the net of an octagonal prism?

9. How many lines of symmetry do the letters in the word 'SYMMETRICAL' have altogether?

10. Which shapes would you need to make a 3D model of an octagonal prism?

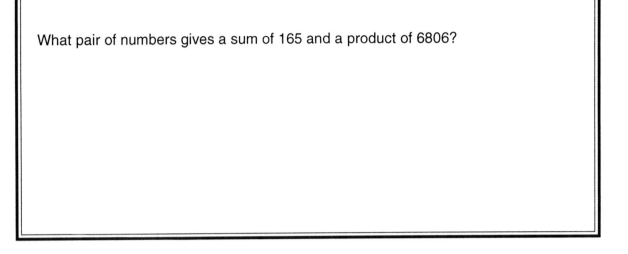

Reasoning about numbers or shapes

Whole class activity

What pair of numbers gives a sum of 165 and a product of 6806?

Explain how you would do this calculation.

36 x 45

Write down an example that matches this statement.

'If you add up all the angles within a triangle it comes to 180º.'

1. What 3 numbers could add up to 0.525?

• •

2. Explain how you would do this calculation.

$$\frac{5}{20} \text{ of } 360$$

• •

3. Write down an example that matches this statement.

'If you multiply a number by a half it makes the number twice as small.'

• •

4. Explain how you can work out the number of days in any number of weeks. Write a formula for it.

• •

5. Write down an example that matches this statement.

'If 13.7 < ☐ < 17.9 then any number between 13.7 and 17.9 can go in the box.'

• •

6. Explain how you would do this calculation.
10,010 − 9625

• •

7. What could the missing numbers be?

**6 ÷ ** = 38

• •

8. Write a formula for the perimeter of a regular polygon where the length of the side is 'ℓ' and the number of sides is 'n' What is the perimeter of a pentagon whose side length is 11cm?

• •

9. Write down an example that matches this statement.

'If you want to multiply a number by 16, multiply it by 4 twice.'

• •

10. Explain how you would do this calculation.

625 ÷ 25

Reasoning about numbers or shapes

1. What 3 numbers could add up to 0.725?

··

2. Explain how you would do this calculation.

   $\dfrac{5}{20}$ of 1360

··

3. Write down an example that matches this statement.

   'If you multiply a number by a quarter it makes the number four times as small.'

··

4. Explain how you can work out the number of months in any number of years. Write a formula for it.

··

5. Write down an example that matches this statement.

   'If 33.7 < _ < 37.9 then any number between 33.7 and 37.9 can go in the box.'

··

6. Explain how you would do this calculation.

   20,010 – 15,625

··

7. What could the missing numbers be?

   ***6 ÷ ** = 328

··

8. Write a formula for the perimeter of a regular polygon where the length of the side is 'ℓ' and the number of sides is 'n'. What is the perimeter of a hexagon whose side length is 12cm?

··

9. Write down an example that matches this statement.

   'If you want to multiply a number by 36, multiply it by 6 twice.

··

10. Explain how you would do this calculation.

    3125 ÷ 25

1. | What 3 numbers could add up to 0.935?

• •

2. | Explain how you would do this calculation.

$\dfrac{5}{20}$ of 3360

• •

3. | Write down an example that matches this statement.

'If you multiply a number by a third it makes the number three times as small.'

• •

4. | Explain how you can work out the number of months in any number of decades. Write a formula for it.

• •

5. | Write down an example that matches this statement.
'If 333.7 < ▢ < 337.9 then any number between 333.7 and 337.9 can go in the box.'

• •

6. | Explain how you would do this calculation.

28,010 − 14,625

• •

7. | What could the missing numbers be?

****6 ÷ *** = 328

• •

8. | Write a formula for the perimeter of a regular polygon where the length of the side is 'ℓ' and the number of sides is 'n' What is the perimeter of an octagon whose side length is 36cm?

• •

9. | Write down an example that matches this statement.

'If you want to multiply a number by 64, multiply it by 8 twice.'

• •

10. | Explain how you would do this calculation.

30125 ÷ 25

This page may be photocopied by the purchasing institution only.

© Catherine Yemm

www.brilliant publications.co.uk

Problem Solving Maths – Year 6   41

Reasoning about numbers or shapes

Whole class activity

This is one quarter of a shape, what could the whole shape look like?

Explain how to work out the area of a right-angled triangle. Write a formula to explain it.

Count the squares you can see.

This page may be photocopied by the purchasing institution only.

© Catherine Yemm

1. How many degrees do the angles of a quadrilateral add up to?

· ·

2. What could the time be if the hands on an analogue clock show an angle of 60º?

· ·

3. How many lines of symmetry does a heptagon have?

· ·

4. Draw a picture with 8 squares using 6 small squares.

· ·

5. Draw and name a 3D shape that has more than 10 edges.

· ·

6. Which shapes would you need to make a 3D model of a pentagonal-based pyramid?

· ·

7. How many different shaped rectangles can you make with 24 cubes?

· ·

8. Name one way in which a hemisphere and a sphere are different.

· ·

9. You have a piece of paper in the shape of a square. What is the smallest number of cuts you have to make to turn the square into a pentagon?

· ·

10. Cut a diamond into quarters. What shapes do you get?

www.brilliantpublications.co.uk

1. How many degrees do the angles of a hexagon add up to?

· ·

2. What could the time be if the hands on an analogue clock show an angle of 90º?

· ·

3. How many lines of symmetry does a decagon have?

· ·

4. Draw a picture with 18 squares using fewer than 12 small squares.

· ·

5. Draw and name a 3D shape that has more than 15 edges.

· ·

6. Which shapes would you need to make a 3D model of a heptagonal-based pyramid?

· ·

7. How many different shaped rectangles can you make with 32 cubes?

· ·

8. Name two ways in which a hemisphere and a sphere are different.

· ·

9. You have a piece of paper in the shape of a square. What is the smallest number of cuts you have to make to turn the square into an octagon?

· ·

10. Cut a diamond into eighths. What shapes do you get?

1. How many degrees do the angles of a nonogon add up to?

· ·

2. What could the time be if the hands on an analogue clock show an angle of 120°?

· ·

3. How many lines of symmetry does a dodecahedron have?

· ·

4. Draw a picture with more than 22 squares using 16 small squares.

· ·

5. Draw and name a 3D shape that has more than 20 edges.

· ·

6. Which shapes would you need to make a 3D model of a decagonal-based pyramid?

· ·

7. How many different shaped rectangles can you make with 64 cubes?

· ·

8. Name three ways in which a hemisphere and a sphere are different.

· ·

9. You have a piece of paper in the shape of a square. What is the smallest number of cuts you have to make to turn the square into a decagon?

· ·

10. Cut a diamond into sixteenths. What shapes do you get?

Reasoning about numbers or shapes

Whole class activity

Write down an example that matches this statement.

'What 5 numbers could give you a total of 11.76?'

If the formula is $3n + 2$ where n is the position of a number, what would the next 5 numbers after 5, 8 be?

Explain how you would do this calculation.

$$\frac{3}{15} \text{ of } 1258$$

Lesson
4a

1. Name 4 three-digit numbers where the product of the digits is 10.

• •

2. Explain how you would do this calculation.

12.5% of 500

• •

3. Write down an example that matches this statement.

'Adding two multiples of 11 will always give an answer that is a multiple of 11.'

• •

4. If one whole number multiplied by a decimal gives an answer of 12.5, what could the number and the decimal be?

• •

5. What five numbers could give you a total of 12.38?

• •

6. Write down an example that matches this statement.

'If you add three consecutive numbers, the answer is always three times as much as the second number.'

• •

7. Explain how you would do this calculation.
60 x 85

• •

8. The size of the zip needed to fix a coat which is *n* centimetres long is *2n + 1*. What length do you need for a coat 30cm long.

• •

9. Write down an example that matches this statement.

'If you have brackets in a calculation, to get the right answer you must work out the answer inside the brackets first.'

• •

10. Explain how you would do this calculation.

634 – 124

Lesson 4b

1. Name 4 three-digit numbers where the product of the digits is 28.

• •

2. Explain how you would do this calculation.

12.5% of 5000

• •

3. Write down an example that matches this statement.

'Adding two multiples of 15 will always give an answer that is a multiple of 15.'

• •

4. If one whole number multiplied by a decimal gives an answer of 212.5, what could the number and the decimal be?

• •

5. What five numbers could give you a total of 2.38?

• •

6. Write down an example that matches this statement.

'If you add five consecutive numbers, the answer is always five times as much as the third number.'
• •

7. Explain how you would do this calculation.

600 x 85

• •

8. The size of the zip needed to fix a coat which is n centimetres long is $2n + 1$. What length zip do you need for a coat 56.4cm long?

• •

9. Write down an example that matches this statement.

'If you have brackets in a calculation to get the right answer you must work out the answer inside the brackets first.'
• •

10. Explain how you would do this calculation.

634.5 – 424.6

© Catherine Yemm

1. Name 4 three-digit numbers where the product of the digits is 24.

· ·

2. Explain how you would do this calculation.

12.5% of 50,000

· ·

3. Write down an example that matches this statement.

'Adding two multiples of 18 will always give an answer that is a multiple of 18.'

· ·

4. If one whole number multiplied by a decimal gives an answer of 2212.5, what could the number and the decimal be?

· ·

5. What five numbers could give you a total of 0.238?

· ·

6. Write down an example that matches this statement.

'If you add seven consecutive numbers, the answer is always seven times as much as the fourth number.'

· ·

7. Explain how you would do this calculation.

6000 x 85

· ·

8. The size of the zip needed to fix a coat which is *n* centimetres long is *2n + 1*. What length zip do you need for a coat 67.25cm long?

· ·

9. Write down an example that matches this statement.

'If you have brackets in a calculation, to get the right answer you must work out the answer inside the brackets first.'

· ·

10. Explain how you would do this calculation.

6234.51 – 4124.63

Reasoning about numbers or shapes

Whole class activity

How many different ways can we split this shape into equal halves?

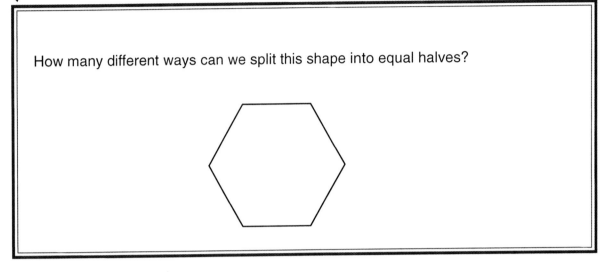

If the length of the side of a heptagon is 11.5cm, how long is the perimeter of the heptagon?

If this is one eighth of a shape, what could the whole shape look like?

Lesson
5a

1. What is the name of a polygon with 9 sides?

2. How many more sides does an octagon have than a quadrilateral?

3. What is the angle between the clock hands if the time is 3 o'clock?

4. If a bicycle wheel has 5 spokes, how many lines of symmetry does it have?

5. Name two shapes that can be put into the group entitled 'Has a curved face'.

6. Draw a shape with 6 lines of symmetry.

7. How many cubes would you need to make a cuboid that is 8 cubes wide, 4 cubes high and 10 cubes long?

8. Join together two squares and two heptagons. How many sides does the new shape have?

9. Cut an octagon into halves. What shapes do you get?

10. This is one half of a shape. What could the shape look like?

Lesson 5b

1. What is the name of a polygon with 11 sides?

• •

2. How many more sides does a nonagon have than a quadrilateral?

• •

3. What is the angle between the clock hands if the time is 2 o'clock?

• •

4. If a bicycle wheel has 10 spokes, how many lines of symmetry does it have?

• •

5. Name three shapes that can be put into the group entitled 'Has a curved face'.

• •

6. Draw a regular shape with 8 lines of symmetry.

• •

7. How many cubes would you need to make a cuboid that is 12 cubes wide, 4 cubes high and 12 cubes long?

• •

8. Join together three squares and two nonagons. How many sides does the new shape have?

• •

9. Cut an octagon into quarters. What shapes do you get?

• •

10. This is one third of a shape. What could the shape look like?

1. What is the name of polygon with 12 sides?

· ·

2. How many more faces does a dodecagon have than an octahedron?

· ·

3. What is the angle between the clock hands if the time is 5 o'clock?

· ·

4. If a bicycle wheel has 20 spokes, how many lines of symmetry does it have?

· ·

5. Name four shapes that can be put into the group entitled 'Has a curved face'.

· ·

6. Draw a regular shape with 10 lines of symmetry.

· ·

7. How many cubes would you need to make a cuboid that is 16 cubes wide, 3 cubes high and 18 cubes long?

· ·

8. Join together two squares, two heptagons and a decagon. How many sides does the new shape have?

· ·

9. Cut a regular octagon into eighths. What shapes do you get?

· ·

10. This is one fifth of a shape. What could the shape look like?

© Catherine Yemm

Write down an example that matches this statement.

'The product of two numbers is not always greater than the sum of two numbers.'

Explain how you would do this calculation.

$522 \div 9$

Write a formula to describe the nth term of this sequence.

5, 10, 15, 20, 25

www.brilliantpublications.co.uk

This page may be photocopied by the purchasing institution only.

54   **Problem Solving Maths – Year 6**

© Catherine Yemm

1. Write down an example that supports this statement.

   'If we divide a decimal by 10, each digit moves one place to the left.'

   •

2. Explain how you would do this calculation.

   22.5% of £2500

   •

3. Write down an example that matches this statement.

   'To divide a number by 50, divide by 100 and multiply by 2.'

   •

4. Name 3 three-digit numbers where the sum of the digits is 12.

   •

5. Explain how you would do this calculation.

   35 x 15

   •

6. Write down a formula to calculate the cost of *n* number of apples at 23p each.

   •

7. Explain how you would do this calculation.

   8569 + 8926

   •

8. One whole number divided by another gives the answer of 19.6. What could the two whole numbers be?

   •

9. Write down an example that supports this statement.

   'A multiple of 20 is always a multiple of 5 and 10.'

   •

10. If 3A + 4B = 84 what numbers could A and B represent?

This page may be photocopied by the purchasing institution only.

© Catherine Yemm

www.brilliantpublications.co.uk

Problem Solving Maths – Year 6   55

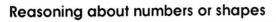

Lesson
6b

1. Write down an example that supports this statement.
'If we divide a decimal by 100, each digit moves two places to the left.'

••

2. Explain how you would do this calculation.

22.5% of £25,000

••

3. Write down an example that matches this statement.

'To divide a number by 25, divide by 100 and multiply by 4.'

••

4. Name 3 three-digit numbers where the sum of the digits is 15.

••

5. Explain how you would do this calculation.

52 x 15

••

6. Write down a formula to calculate the cost of _n_ number of apples at 32p each.

••

7. Explain how you would do this calculation.

13 569 + 13 926

••

8. One whole number divided by another gives the answer of 39.6. What could the two whole numbers be?

••

9. Write down an example that supports this statement.

'A multiple of 50 is always a multiple of 25 and 10.'

••

10. If 3A + 4B = 127 what numbers could A and B represent?

1. Write down an example that supports this statement.

   'If we divide a decimal by 1000, each digit moves three places to the left.'

   •

2. Explain how you would do this calculation.

   22.5% of £250,000

   •

3. Write down an example that matches this statement.

   'To divide a number by 20, divide by 100 and multiply by 5.'

   •

4. Name 3 three-digit numbers where the sum of the digits is 18.

   •

5. Explain how you would do this calculation.

   64 x 15

   •

6. Write down a formula to calculate the cost of n number of apples at 64p each.

   •

7. Explain how you would do this calculation.

   21 569 + 21 926

   •

8. One whole number divided by another gives the answer of 59.6. What could the two whole numbers be?

   •

9. Write down an example that supports this statement.

   'A multiple of 100 is always a multiple of 25 and 10.'

   •

10. If 3A + 4B = 1270 what numbers could A and B represent?

Problems involving 'real life', money or measures

Whole class activity

Last year the local library had 8560 members. This year 4436 new members have joined. How many members does the library have now?

Miss Lynn needs 7.5kg of modelling clay for her class to all make masks. She has 2750g. How many more kilograms does she need to buy?

Amanda competed in a triathlon. She swam 5km and 33 metres, cycled 4.2km, and ran 15.8 kilometres. How many metres did she travel altogether?

Lesson 1a

1. If there are 1.49 euros to £1, how much will it cost in £s to go on a fishing trip around Spain if the trip costs 198 euros?

• •

2. A small pot of glue holds 400ml of glue. If a large container of glue holds 8.4 litres, how many small pots will a large container of glue fill?

• •

3. Jade and Shen are asked to share 360 crayons into 18 bags. How many crayons will be in each bag? Ellie picks up 8 bags. How many crayons is she holding?

• •

4. Mr Shrub's breadmaker cooks different weights of bread for different amounts of time. If it cooks the bread for 20 minutes for every 100g of flour, how long will a loaf of bread made with $\frac{1}{2}$ a kilogram of flour be cooked for?

• •

5. There are 145 books in the library. Two-fifths of them are out on loan. How many books are still in the library?

• •

6. What is the area of the school kitchen floor if it is 20 tiles wide and 30 tiles in length and each tile measures 10cm x 10cm?

• •

7. The cost of a new bike without VAT is £35.50. What will the total price be once the 17.5% VAT has been added?

• •

8. The local school is putting on a fashion show for charity. 673 tickets have been sold at £4.25 each. How much money has been raised for charity?

• •

9. The school playground has an area of 80m². The playground is not a rectangle or a square shape. What shape could it be? What could the perimeter be?

• •

10. Mr Davies bought 42.3kg of soil to put on his garden. He has 1460g of soil left. How much has he put on his garden?

Lesson 1b

1. If there are 1.49 euros to £1, how much will it cost in £s to go on a fishing trip around Spain if the trip costs 398 euros?

....

2. A small pot of glue holds 400ml of glue. If a large container of glue holds 16.4 litres, how many small pots will a large container of glue fill?

....

3. Jade and Shen are asked to share 1080 crayons into 18 bags. How many crayons will be in each bag? Ellie picks up 8 bags. How many crayons is she holding?

....

4. Mr Shrub's breadmaker cooks different weights of bread for different lengths of time. If it cooks the bread for 20 minutes for every 100g of flour, how long will a loaf of bread made with $\frac{1}{4}$ of a kilogram of flour be cooked for?

....

5. There are 345 books in the library. Two-fifths of them are out on loan. How many books are still in the library?

....

6. What is the area of the school kitchen if it is 30 tiles wide and 40 tiles in length and each tile measures 10cm x 10cm?

....

7. The cost of a new bike without VAT is £75.95. What will the total price be once the 17.5% VAT has been added?

....

8. The local school is putting on a fashion show for charity. 1673 tickets have been sold at £4.25 each. How much money has been raised for the charity?

....

9. The school playground has an area of 120m². The playground is not a rectangle or a square shape. What shape could it be? What could the perimeter be?

....

10. Mr Davies bought 42.3kg of soil to put on his garden. He has 8460g of soil left. How much has he put on his garden?

1. If there are 1.49 euros to £1, how much will it cost in £s to go on a fishing trip around Spain if the trip costs 498 euros?

2. A small pot of glue holds 400ml of glue. If a large container of glue holds 32.4 litres, how many small pots will a large container of glue fill?

3. Jade and Shen are asked to share 2214 crayons into 18 bags. How many crayons will be in each bag? Ellie picks up 8 bags. How many crayons is she holding?

4. Mr Shrub's breadmaker cooks different weights of bread for different lengths of time. If it cooks the bread for 20 minutes for every 100g of flour, how long will a loaf of bread made with $\frac{2}{5}$ of a kilogram of flour be cooked for?

5. There are 545 books in the library. Two-fifths of them are out on loan. How many books are still in the library?

6. What is the area of the school kitchen if it is 40 tiles wide and 50 tiles in length and each tile measures 10cm x 10cm?

7. The cost of a new bike without VAT is £175.95. What will the total price be once the 17.5% VAT has been added?

8. The local school is putting on a fashion show for charity. 2673 tickets have been sold at £4.25 each. How much money has been raised for the charity?

9. The school playground has an area of 220m². The playground is not a rectangle or a square shape. What shape could it be? What could the perimeter be?

10. Mr Davies bought 62.3kg of soil to put on his garden. He has 2460g of soil left. How much has he put on his garden?

Problems involving 'real life', money or measures

Whole class activity

Rosie has made a box that measures 20cm wide, 5cm in length and 30cm in height. What area of wrapping paper would she need to cover the box?

I think of a number. I add 4.8 and multiply by 6. The answer is 34.2. What was my number?

The time in New York is 5 hours behind the time in London. If it is 12:17pm. in London what time will it be in New York in $8\frac{1}{2}$ hours' time?

1. The local park is 21 metres long and 10 metres wide. The council want to put a spongy surface down on a third of the park. What will be the size of the area still covered by grass?

   •

2. Class 6 have a sink that holds 4.2 litres of water. Class 5 have a sink that holds two and a half times as much. How much water does class 5's sink hold?

   •

3. Packets of crisps in the local shop are 16p each. They are packed in boxes of 50. If the shopkeeper sells 2 boxes how much money will he have made?

   •

4. The school has sold 138 tickets to its concert. The caretaker can fit 45 chairs in one row. If the caretaker only puts out full rows of chairs, how many full rows of chairs will there be? How many seats will not be used?

   •

5. Anna wants to buy a new dress. The dress was £30.00 but it is in the sale with 15% off. How much will the dress be?

   •

6. Bethan's bag of sweets weighs 2lb. Caitlin's bag of sweets weighs 2kg. Who has more sweets? By how much?

   •

7. Mrs Nelson is cutting up a long piece of rope so the children in her class can all have a piece to practise tying knots. Her rope is 10.20 metres and there are 34 children in the class. How much will they have each?

   •

8. Victoria has just been on holiday to Australia. She took £45 to spend but she only spent £12.50. If the exchange rate is £1 to 2.5 Australian dollars how many Australian dollars has she brought back?

   •

9. Kedar is helping to organize the school sports cupboard. He can fit 32 tennis balls in a box. The school has 256 tennis balls. How many boxes does he need?

   •

10. Mr Jessop, the school caretaker, orders some fencing to go around the school field. The width of the field is 24.6 metres. The length is double that. He also needs to order an extra 5% in case he damages some when he is making the fence. How much fencing does Mr Jessop need to order?

Lesson 2b

1. The local park is 45 metres long and 20 metres wide. The council want to put a spongy surface down on a third of the park. What will be the size of the area still covered by grass?

2. Class 6 have a sink that holds 4.2 litres of water. Class 5 have a sink that holds three and a half times as much. How much water does class 5's sink hold?

3. Packets of crisps in the local shop are 26p each. They are packed in boxes of 50. If the shopkeeper sells 2 boxes how much money will he have made?

4. The school has sold 438 tickets to its concert. The caretaker can fit 45 chairs in one row. If the caretaker only puts out full rows of chairs, how many full rows of chairs will there be? How many seats will not be used?

5. Anna wants to buy a new dress. The dress was £23.00 but it is in the sale with 15% off. How much will the dress be?

6. Bethan's bag of sweets weighs 3lb. Caitlin's bag of sweets weighs 3kg. Who has more sweets? By how much?

7. Mrs Nelson is cutting up a long piece of rope so the children in her class can all have a piece to practise tying knots. Her rope is 11.05 metres and there are 34 children in the class. How much will they have each?

8. Victoria has just been on holiday to Australia. She took £55 to spend but she only spent £22.50. If the exchange rate is £1 to 2.5 Australian dollars how many Australian dollars has she brought back?

9. Kedar is helping to organize the school sports cupboard. He can fit 32 tennis balls in a box. The school has 544 tennis balls. How many boxes does he need?

10. Mr Jessop, the school caretaker, orders some fencing to go around the school field. The width of the field is 34.6 metres. The length is double this. He also needs to order an extra 5% in case he damages some when he is making the fence. How much fencing does Mr Jessop need to order?

1. The local park is 65 metres long and 20 metres wide. The council want to put a spongy surface down on a third of the park. What will be the size of the area still covered by grass?

2. Class 6 have a sink that holds 8.2 litres of water. Class 5 have a sink that holds two and a half times as much. How much water does class 5's sink hold?

3. Packets of crisps in the local shop are 46p each. They are packed in boxes of 50. If the shopkeeper sells 2 boxes how much money will he have made?

4. The school has sold 738 tickets to its concert. The caretaker can fit 45 chairs in one row. If the caretaker only puts out full rows of chairs, how many full rows of chairs will there be? How many seats will not be used?

5. Anna wants to buy a new dress. The dress was £43.00 but it is in the sale with 15% off. How much will the dress be?

6. Bethan's bag of sweets weighs 4lb. Caitlin's bag of sweets weighs 4kg. Who has more sweets? By how much?

7. Mrs Nelson is cutting up a long piece of rope so the children in her class can all have a piece to practise tying knots. Her rope is 20.40 metres and there are 34 children in the class. How much will they have each?

8. Victoria has just been on holiday to Australia. She took £75 to spend but she only spent £42.50. If the exchange rate is £1 to 2.5 Australian dollars how many Australian dollars has she brought back?

9. Kedar is helping to organize the school sports cupboard. He can fit 32 tennis balls in a box. The school has 960 tennis balls. How many boxes does he need?

10. Mr Jessop, the school caretaker, orders some fencing to go around the school field. The width of the field is 54.6 metres. The length is double this. He also needs to order an extra 5% in case he damages some when he is making the fence. How much fencing does Mr Jessop need to order?

Problems involving 'real life', money or measures

The local sports centre's costs are £1.25 for swimming, £1.85 for badminton and £1.95 for tennis. Gina does two activities and has £6.20 change from £10. Which 2 activities does she do?

Classes 3, 4, 5 and 6 are going on a school trip together. There are 38 children in each class, and 6 adults per class are going to accompany them. A school bus has 52 seats. How many buses will be needed to take the children? How many empty seats will there be?

There are 80 Indian rupees to the pound. What would be the price in £s of a house in India costing 3,600,000 rupees?

1. At Hannah's birthday party her mum gave a quarter of the cake to the adults and four-tenths of the cake to the boys. How much of the cake was left for the girls?

2. Mrs Thomson gives her class 288 raffle tickets and asks them to make them into books of 18. How many books of raffle tickets will they have to sell at the school open day?

3. A local turf company has donated 24m² of turf to the local nursery. How long and wide would their mud patch be for them to be able to use all of the turf?

4. Jaya sent a birthday present to her friend. One of the two presents inside the parcel weighed 1.2kg, the other weighed 4.35kg. The paper to wrap the parcel weighed 500g and the string weighed 20g. How much did the parcel weigh altogether?

5. The local football ground usually seats 156 spectators. The council has put up some extra stands so it now seats 25% more fans. How many people will the ground seat now?

6. The children in class 6 are having a snail race. There are 3 snails. Snail A travels 13.1cm, Snail B travels 14.7cm, Snail C travels 19.6cm. How many more centimetres would they have to travel if they were to travel a metre between them?

7. It starts raining on 1st September and the rainfall collected measures 25ml. If it rains every day for a fortnight and the rain collected increases by 2ml a day, how much rain would have been collected by the end of the fortnight?

8. Joel has 3 favourite films. The first is 1hr and 5 minutes long, the second is 1 minute longer than the first, and the third is 10 minutes longer than the second. He starts watching the films at 3.00 pm and watches them one after another with a 10 minute break in between. What time does he finish watching the films?

9. In the local pet shop there are 48 animals. Two-thirds are mammals, 4 are reptiles and a quarter are birds. How many animals are not mammals, reptiles or birds?

10. The athletics track at the local sports ground has a perimeter of 120m. If Joseph runs around this 17 times how many kilometres will he have run?

Lesson 3b

1. At Hannah's birthday party her mum gave a quarter of the cake to the adults, four-tenths of the cake to the boys and two-eighths of the cake to the girls. How much of the cake was left?

2. Mrs Thomson gives her class 576 raffle tickets and asks them to make them into books of 18. How many books of raffle tickets will they have to sell at the school open day?

3. A local turf company has donated 36m² of turf to the local nursery. How long and wide would their mud patch be for them to be able to use all of the turf?

4. Jaya sent a birthday present to her friend. One of the two presents inside the parcel weighed 3.2kg, the other weighed 6.35kg. The paper to wrap the parcel weighed 500g and the string weighed 20g. How much did the parcel weigh altogether?

5. The local football ground usually seats 256 spectators. The council has put up some extra stands so it now seats 25% more fans. How many people will the ground seat now?

6. The children in class 6 are having a snail race. There are 3 snails. Snail A travels 30.1cm, Snail B travels 24.7cm, Snail C travels 29.6cm. How many more centimetres would they have to travel if they were to travel a metre between them?

7. It starts raining on 1st September and the rainfall collected is 25ml. If it rains every day for a fortnight and the rain collected increases by 4ml a day, how much rain will have been collected by the end of the fortnight?

8. Joel has 3 favourite films. The first is 1hr and 15 minutes long, the second is 2 minutes longer than the first, and the third is 20 minutes longer than the second. He starts watching the films at 3.00 pm and watches them one after another with a 10 minute break in between. What time does he finish watching the films?

9. In the local pet shop there are 108 animals. Two-thirds are mammals, 4 are reptiles, a quarter are birds. How many animals are not mammals, reptiles or birds?

10. The athletics track at the local sports ground has a perimeter of 120m. If Joseph runs around this 19 times how many kilometres will he have run?

1. At Hannah's birthday party her mum gave three-twelfths of the cake to the adults, four-tenths of the cake to the boys and two-eighths of the cake to the girls. How much of the cake was left?

   •

2. Mrs Thomson gives her class 1152 raffle tickets and asks them make them into books of 18. How many books of raffle tickets will they have to sell at the school open day?

   •

3. A local turf company has donated 54m² of turf to the local nursery. How long and wide would their mud patch be for them to be able to use all of the turf?

   •

4. Jaya sent a birthday present to her friend. One of the two presents inside the parcel weighed 13.2kg, the other weighed 16.35kg. The paper to wrap the parcel weighed 500g and the string weighed 20g. How much did the parcel weigh altogether?

   •

5. The local football ground usually seats 512 spectators. The council has put up some extra stands so it now seats 25% more fans. How many people will the ground seat now?

   •

6. The children in class 6 are having a snail race. There are 3 snails. Snail A travels 40.1cm, Snail B travels 44.7cm, Snail C travels 49.6cm. How many more centimetres would they have to travel if they were to travel two metres between them?

   •

7. It starts raining on 1st September and the rainfall collected is 25ml. If it rains every day for a fortnight and the rain collected increases by 7ml a day, how much rain would have been collected by the end of the fortnight?

   •

8. Joel has 3 favourite films. The first is 1hr and 27 minutes long, the second is 4 minutes longer than the first, and the third is 14 minutes longer than the second. He starts watching the films at 3.00 pm and watches them one after another with an 18 minute break in between. What time does he finish watching the films?

   •

9. In the local pet shop there are 204 animals. Two-thirds are mammals, 4 are reptiles, a quarter are birds. How many animals are not mammals, reptiles or birds?

   •

10. The athletics track at the local sports ground has a perimeter of 120m. If Joseph runs around this 21 times how many kilometres will he have run?

Problems involving 'real life', money or measures

Whole class activity

Jamie's dad is selling his car. The man who is helping him sell it takes 3% of the sale price as payment for his work. If Jamie's dad sells his car for £8275, how much money will the seller make?

The Year 6 classroom is made up of two areas. The wet area where the sink is measures 3 metres by 6 metres and the carpeted area measures 8 metres by 12 metres. What is the area of the classroom?

The local high school takes in children from seven local primary schools. This year the intake is 467. What is the average number of children coming from each primary school?

1. The school canteen has ordered 1.4 litres of squash. To make one cup of squash they need to mix 50ml of squash with 150ml of water. How many cups of squash can they make? How many litres of water will they use?

2. The school has won a competition and has been awarded £246 to spend on classroom resources. There are 6 classes in the school. How much will each class have to spend?

3. The school library has 6 aisles. Each aisle has 8 shelves and each shelf holds 20 books. How many books are in the library?

4. In France a book costs 5.88 euros. The same book in Great Britain costs £4. What is the exchange rate?

5. Sarah is practising her running. She starts off on Monday by being able to run 400 metres in 2 minutes and 40 seconds. Each day she increases her speed by ten seconds. How fast will she be running by Friday?

6. If there is 2.54cm to an inch, how many centimetres long are the tables in the classroom if they measure 15 inches in length?

7. A box of books weighs 36kg. The books inside are of similar size and shape. If there are 12 books inside the box what does the average book weigh?

8. Class 6 carried out a survey and found that their class of 35 pupils watched 188 hours of television per week altogether. What is the average length of time in a week each pupil spends watching television?

9. I think of a number and add 7.25 and multiply by 5. The answer is 88.75. What was my number?

10. The local park is 14m wide and 12m long. The man who cuts the grass uses a mower that cuts 2m² of grass in 3 seconds. How long will it take him to cut the grass in the park?

Lesson 4b

1. The school canteen has ordered 2.4 litres of squash. To make one cup of squash they need to mix 50ml of squash with 150ml of water. How many cups of squash can they make? How many litres of water will they use?

• •

2. The school has won a competition and has been awarded £546 to spend on classroom resources. There are 6 classes in the school. How much will each class have to spend?

• •

3. The school library has 6 aisles. Each aisle has 18 shelves and each shelf holds 30 books. How many books are in the library?

• •

4. In France a book costs 13.59 euros. The same book in Great Britain costs £9. What is the exchange rate?

• •

5. Sarah is practising her running. She starts off on Monday by being able to run 400 metres in 2 minutes and 43 seconds. Each day she increases her speed by ten seconds. How fast will she be running by Friday?

• •

6. If there is 2.54cm to an inch, how many centimetres long are the tables in the classroom if they measure 22 inches in length?

• •

7. A box of books weighs 34kg. The books inside are of similar size and shape. If there are 25 books inside the box what does the average book weigh?

• •

8. Class 6 carried out a survey and found that their class of 35 pupils watched 288 hours of television per week altogether. What is the average length of time in a week each pupil spends watching television?

• •

9. I think of a number and add 7.25 and multiply by 5. The answer is 108.75. What was my number?

• •

10. The local park is 24m wide and 32m long. The man who cuts the grass uses a mower that cuts 2m² of grass in 3 seconds. How long will it take him to cut the grass in the park?

Lesson
4c

1. The school canteen has ordered 4.8 litres of squash. To make one cup of squash they need to mix 50ml of squash with 150ml of water. How many cups of squash can they make? How many litres of water will they use?

....

2. The school has won a competition and has been awarded £846 to spend on classroom resources. There are 6 classes in the school. How much will each class have to spend?

....

3. The school library has 12 aisles. Each aisle has 18 shelves and each shelf holds 30 books. How many books are in the library?

....

4. In France a book costs 16.56 euros. The same book in Great Britain costs £12. What is the exchange rate?

....

5. Sarah is practising her running. She starts off on Monday by being able to run 400 metres in 3 minutes and 23 seconds. Each day she increases her speed by eight seconds. How fast will she be running by Friday?

....

6. If there is 2.54cm to an inch, how many centimetres long are the tables in the classroom if they measure 42 inches in length?

....

7. A box of books weighs 64kg. The books inside are of similar size and shape. If there are 25 books inside the box what does the average book weigh?

....

8. Class 6 carried out a survey and found that their class of 35 pupils watched 488 hours of television per week altogether. What is the average length of time in a week each pupil spends watching television?

....

9. I think of a number and add 7.25 and multiply by 15. The answer is 255. What was my number?

....

10. The local park is 34m wide and 42m long. The man who cuts the grass uses a mower that cuts 2m² of grass in 3 seconds. How long will it take him to cut the grass in the park?

© Catherine Yemm

Problems involving 'real life', money or measures

Whole class activity

The prices of 4 items in a shop are £111.36, £26.34, £145.56 and £82. How much would you spend if you bought one of each of these items in a half price sale?

There are 8 houses in Thomas's street. To paint his house Thomas's dad needs 352 litres of paint. If he was to paint all the houses in the street how much paint would he need?

At last year's school concert 240 tickets were sold. This year the head teacher is expecting this number to rise by 5%. How many tickets should be printed?

Lesson
5a

1. The children in class 6 have been carrying out a survey on sleeping. Between them they sleep for 1529.5 hours in a week. If there are 23 children in the class what is the average amount of sleep the children get each night?

· ·

2. Last year the school spent £1568 on books for the children. This year a local charity has donated £856 to buy books. How much will the school need to spend this year if it wants to buy the same value of books for the children as it did the year before?

· ·

3. George has saved £87.40 in his bank account. How much will he have left if he spends a half of his savings on a new scooter?

· ·

4. There are 16oz in a lb and 14lbs in a stone. How much does Jacob weigh in stones and lbs if he weighs 672 oz?

· ·

5. The head teacher is buying some furniture for his office. He buys a desk for £23.75, a filing cabinet for £42.60 and a set of shelves for £18.90. How much does he spend?

· ·

6. Tracey's sunflower was 3.2 metres tall but 45 centimetres was snapped off the top in the wind. How tall is it now?

· ·

7. It is Tuesday 24th March, 09:54 in London. What date and time is it in Sydney if they are 10 hours ahead of London?

· ·

8. The local bus travels from the village to the nearest town and back 4 times a day. It is 8.4 miles from the village to the town. How many miles a day does the bus driver travel?

· ·

9. The school cook is making cakes and is baking them on a baking sheet in rows. In each row are 4 chocolate cakes and 5 strawberry cakes. If she bakes 24 chocolate cakes how many strawberry cakes does she bake?

· ·

10. Dafydd buys a new book in the sale. The shopkeeper gives him 20% off and he pays £8.60. How much did the book cost before the discount?

Lesson 5b

1. The children in class 6 have been carrying out a survey on sleeping. Between them they sleep for 1851.5 hours in a week. If there are 23 children in the class what is the average amount of sleep the children get each night?

2. Last year the school spent £2568 on books for the children. This year a local charity has donated £856 to buy books. How much will the school need to spend this year if it wants to buy the same value of books for the children as it did the year before?

3. George has saved £127.40 in his bank account. How much will he have left if he spends a quarter of his savings on a new scooter?

4. There are 16oz in a lb and 14lbs in a stone. How much does Jacob weigh in stones and lbs if he weighs 896oz?

5. The head teacher is buying some furniture for his office. He buys a desk for £123.75, a filing cabinet for £42.60 and a set of shelves for £18.90. How much does he spend?

6. Tracey's sunflower was 4.2 metres tall but 85 centimetres was snapped off the top in the wind. How tall is it now?

7. It is Tuesday 24th March, 16:54 in London. What date and time is it in Sydney if they are 10 hours ahead of London?

8. The local bus travels from the village to the nearest town and back 12 times a day. It is 8.4 miles from the village to the town. How many miles a day does the bus driver travel?

9. The school cook is making cakes and is baking them on a baking sheet in rows. In each row are 4 chocolate cakes and 5 strawberry cakes. If she bakes 48 chocolate cakes how many strawberry cakes does she bake?

10. Dafydd buys a new book in the sale. The shopkeeper gives him 20% off and he pays £11.60. How much did the book cost before the discount?

1. The children in class 6 have been carrying out a survey on sleeping. Between them they sleep for 1368.5 hours in a week. If there are 23 children in the class what is the average amount of sleep the children get each night?

2. Last year the school spent £4568 on books for the children. This year a local charity has donated £1856 to buy books. How much will the school need to spend this year if it wants to buy the same value of books for the children as it did the year before?

3. George has saved £427.40 in his bank account. How much will he have left if he spends a quarter of his savings on a new scooter?

4. There are 16 ounces in a pound and 14 pounds in a stone. How much does Jacob weigh in stones and pounds if he weighs 1344 ounces?

5. The head teacher is buying some furniture for his office. He buys a desk for £223.75, a filing cabinet for £82.60 and a set of shelves for £78.90. How much does he spend?

6. Tracey's sunflower was 3.27 metres tall but 189 centimetres was snapped off the top in the wind. How tall is it now?

7. It is Tuesday 24th March, 16:54 and 56 seconds in London. What date and time is it in Sydney if they are 10 hours ahead of London?

8. The local bus travels from the village to the nearest town and back 18 times a day. It is 8.4 miles from the village to the town. How many miles a day does the bus driver travel?

9. The school cook is making cakes and is baking them on a baking sheet in rows. In each row are 4 chocolate cakes and 5 strawberry cakes. If she bakes 64 chocolate cakes, how many strawberry cakes does she bake?

10. Dafydd buys a new book in the sale. The shopkeeper gives him 20% off and he pays £28.60. How much did the book cost before the discount?

This page may be photocopied by the purchasing institution only.

www.brilliantpublications.co.uk

© Catherine Yemm

Maths Problem Solving – Year 6   77

Problems involving 'real life', money or measures

Whole class activity

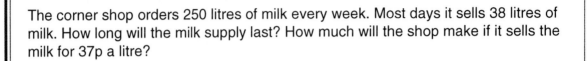

The corner shop orders 250 litres of milk every week. Most days it sells 38 litres of milk. How long will the milk supply last? How much will the shop make if it sells the milk for 37p a litre?

Here is the recipe for making a casserole. Change the weights to approximate metric units. 1oz is approximately 25g and 1 pint is approximateley 570ml.

6oz carrots
5oz potatoes
1lb 3oz of chicken
1 pint of chicken stock

Caleb saves £1.75 a week. How much will he have saved after 2 years? Will he have enough to buy a skateboard costing £132 + VAT at 17.5%?

1. Mr James recently ran a local marathon. The marathon was 26 miles long. He walked for 6.5 miles and jogged for 13 miles and ran the rest of the way. What fraction of the race did he walk, jog and run?

2. I think of a number. I subtract 25 and divide by 5. The answer is 23. What was my number?

3. The local shop has these offers: 3 pairs of socks for £3.30, 5 hats for £12.50, 4 pairs of gloves for £6.40. How much does one of each of these things cost?

4. The local farmer owns 2 sheds on his land One covers an area of 84m² and the other one covers an area of 40m². If he knocks them down to build one new shed with the same area what could the dimensions be?

5. Bobby is having a new bike for his birthday. It is 5.30 pm on 12th September. Bobby's bike is arriving at 9.00 pm on 13th September. How many hours does he have to wait?

6. Sakti cycles his bike at 24.4km an hour. It takes him 30 minutes to cycle to his grandma's house. How many metres away does his grandma live?

7. Joshan has £10 to spend. The exchange rate is £1 to 1.76 US dollars. How many drinks costing 2 dollars can he buy?

8. The junior school in the village uses 4 and a quarter times as much water as the infant school. If the infant school uses 124 litres of water a month, how much does the junior school use?

9. A box of paper clips holds 205. The school secretary orders 9 boxes. How many paper clips will the school have altogether?

10. At the camping shop Catherine buys a new tent worth £83.45 and a new sleeping bag worth £12.50. What will she have to pay if the shopkeeper adds VAT at 17.5% to the total?

Problems involving 'real life', money or measures

1. Mr James recently ran a local marathon. The marathon was 26 miles long. He walked for 6.5 miles and jogged for 16.25 miles and ran the rest of the way. What fraction of the race did he walk, jog and run?

• •

2. I think of a number. I subtract 25 and divide by 2.5. The answer is 24. What was my number?

• •

3. The local shop has these offers: 3 pairs of socks for £4.35, 5 hats for £11.25, 4 pairs of gloves for £6.40. How much does one of each of these things cost?

• •

4. The local farmer has 2 sheds on his land. One covers an area of 112m² and the other one covers an area of 80m². If he knocks them down to build one new shed with the same area what could the dimensions be?

• •

5. Bobby is having a new bike for his birthday. It is 5.30 pm on 12th September. Bobby's bike is arriving at 9.00 pm on 18th September. How many hours does he have to wait?

• •

6. Sakti cycles his bike at 24.4km an hour. It takes him 45 minutes to cycle to his grandma's house. How many metres away does his grandma live?

• •

7. Joshan has £20 to spend. The exchange rate is £1 to 1.76 US dollars. How many drinks costing 2 dollars can he buy?

• •

8. The junior school in the village uses 4 and a quarter times as much water as the infant school. If the infant school uses 254 litres of water a month, how much does the junior school use?

• •

9. A box of paper clips holds 205. The school secretary orders 17 boxes. How many paper clips will the school have altogether?

• •

10. At the camping shop Catherine buys a new tent worth £133.45 and a new sleeping bag worth £28.50. What will she have to pay if the shopkeeper adds VAT at 17.5% to the total?

Lesson
6C

1. Mr James recently ran a local marathon. The marathon was 26 miles long. He walked for 3.05 miles and jogged for 16.45 miles and ran the rest of the way. What fraction of the race did he walk, jog and run?

• •

2. I think of a number. I subtract 25 and divide by 4.5. The answer is 72. What was my number?

• •

3. The local shop has these offers: 3 pairs of socks for £13.50, 5 hats for £18.75, 4 pairs of gloves for £6.40. How much does one of each of these things cost?

• •

4. The local farmer has 2 sheds on his land. One covers an area of 148m^2 and the other one covers an area of 120m^2. If he knocks them down to build one new shed with the same area what could the dimensions be?

• •

5. Bobby is having a new bike for his birthday. It is 5.30p.m. on 12th September. Bobby's bike is arriving at 9.00 pm on 21st September. How many hours does he have to wait?

• •

6. Sakti cycles his bike at 24.4km an hour. It takes him 90 minutes to cycle to his grandma's house. How many metres away does his grandma live?

• •

7. Joshan has £35 to spend. The exchange rate is £1 to 1.76 US dollars. How many drinks costing 2 dollars can he buy?

• •

8. The junior school in the village uses 6 and a quarter times as much water as the infant school. If the infant school uses 254 litres of water a month, how much does the junior school use?

• •

9. A box of paper clips holds 205. The school secretary orders 21 boxes. How many paper clips will the school have altogether?

• •

10. At the camping shop Catherine buys a new tent worth £233.45 and a new sleeping bag worth £78.50. What will she have to pay if the shopkeeper adds VAT at 17.5% to the total?

Answers

Making decisions

Lesson 1 (page 10)
A: 9 hours and 8 minutes; B: £3037.50;
C: divide

Lessons 1a–1c (pages 11–13)

Q	1a	1b	1c
1	story	story	story
2	156	312	432
3	50	45	50
4	£113.73	£313.73	£523.73
5	625ml	812.5ml	1312.5ml
6	0.89	1.66	4.76
7	£4.50	£11.75	£8.42
8	22, no	37, 11 over	72, 10 over
9	18:55	18:51	21:11
10	10 inf. 20 jun.	18 inf. 36 jun.	28 inf. 56 jun.

Lesson 2 (page 14)
A: 2550; B: £247; C: 32.96

Lessons 2a–2c (pages 15–17)

Q	2a	2b	2c
1	139.7mm	266.7mm	520.7mm
2	multiply	multiply	multiply
3	7	11	25
4	story	story	story
5	711 litres	2111 litres	21111 litres
6	850	1724	3424
7	20	20	50
8	317	960	2050
9	8:55	09:25	09:25
10	110.5m	180.5m	1080.5m

Lesson 3 (page 18)
A:1600g; B: 112.5m^2; C: 99 min. 45secs.

Lessons 3a–3c (pages 19–21)

Q	3a	3b	3c
1	18:15	19:15	19:35
2	7	9	13
3	add	add	add
4	254	354	354
5	£422	£1833	£3033
6	359	1359	2641
7	357	2607	6357
8	story	story	story
9	13.2m	21.2m	41.2m
10	228	678	1128

Lesson 4 (page 22)
A: 11820 seconds; B: 1.63m; C: 280.

Lessons 4a–4c (pages 23–25)

Q	4a	4b	4c
1	12	24	48
2	£1.55	£9.55	£19.55
3	1050.5	1350.5	1810.5
4	£3300	£8400	£15,120
5	3 min. 58 sec.	14 min. 18 sec.	19 min. 38 sec.
6	82.4	82.4	92.4
7	minus	minus	minus
8	8	14	37
9	350m	550m	950m
10	story	story	story

Lesson 5 (page26)

A: 384;

B: Tara – 1423, Tilak – 1111, Shauna – 880, Total = 3414

C: 36.

Lessons 5a–5c (pages 27–29)

Q	5a	5b	5c
1	132	252	372
2	20 min.	60 min.	120 min.
3	44	56	80
4	story	story	story
5	365	1365	1365
6	837	2837	7837
7	divide	divide	divide
8	21.5kg	33.5kg	58.5kg
9	421	800	10800
10	224	448	728

Lesson 6 (page 30)

A: 1 hour, 17 minutes;

B: 18; C: 1305.

Lessons 6a–6c (pages 31–33)

Q	6a	6b	6c
1	Martin's by £11.25	Martin's by £10.25	Martin's by £38.25
2	324 inch	540 inch	756 inch
3	multiply	multiply	multiply
4	£28	£34	£48
5	730cm	1530cm	2730cm
6	any	any	any
7	637	1637	2637
8	story	story	story
9	75	110	135
10	187	378	522

Reasoning about numbers or shapes

Lesson 1 (page 34)

A: 5 o'clock and 7 o'clock;

B: 81; C: 7.

Lessons 1a–1c (pages 35–37)

1a heptagonal prism etc.

1b decahedron, octagonal prism etc.

1c dodecahedron, nonagonal prism etc.

Q	1a	1b	1c
2	16	32	64
3a	2 quadilaterals or 2 triangles		
3b	1 rectangle, 2 triangles or 2 quadrilaterals etc		
3c	4 quadrilaterals or 4 triangles etc		
4	any angle that adds up to		
	180°	540°	1080°
5	4096	8000	15 635
6	children to draw shape with		
	9 sides	10 sides	12 sides
7	540°/108°ea	720°/120°ea	1286°/128°ea
8	yes	no	no
9	6	9	9
10	2 hexagons 1 rectangle	2 heptagons 1 rectangle	2 octagons 1 rectangle

Lesson 2 (page 38)

A: 82, 83; B: show workings out for multiplication; C: show examples

Lessons 2a–2c (pages 39–41)

Q	2a	2b	2c
1	any	any	any
2	show workings for fractions		
3	show workings and example		
4	$7 \times n$(wks) =$7n$	$12 \times n$(yrs)=$12n$	$12 \times 12 \times n$ (decades)=$144n$
5	examples	examples	examples
6	show workings for subtraction		
7	$456 \div 12$ $646 \div 17$ etc	$3936 \div 12$ $5576 \div 17$ etc	$10496 \div 32$ $12136 \div 37$

Lessons 2a–2c (pages 39–41)

Q	2a	2b	2c
8	*l x n* 55cm	*l x n* 72cm	*l x n* 288cm
9	examples	examples	examples
10	show workings for division		

Lesson 3 (page 42)

A: draw shapes; B: $\frac{h \times b}{2}$; C: 64

Lessons 3a–3c (pages 43–45)

Q	3a	3b	3c
1	360°	720°	1260°
2	2 & 10 o'clock	3 & 9 o'clock	4 & 8 o'clock
3	7	10	12
4	children to draw using squares		
5	various	various	various
6	1 pentagon 5 triangles	1 heptagon 7 triangles	1 decogon 10 triangles
7	24 x 1,12 x 2 8 x 3, 4 x 6	32 x 1, 16 x 2 8 x 4	64 x 1, 32 x 2 16 x 4, 8 x 8
8	a hemisphere... has 2 faces	has a flat face	has a circular face
9	1	4	6
10	triangles	triangles	triangles

Lesson 4 (page 46)

A: any; B: 11, 14, 17, 20, 23; C: discussion about workings of fractions.

Lessons 4a–4c (pages 47–49)

Q	4a	4b	4c
1	512,125, 152, 521	471, 174, 722, 272	262, 622, 342, 243
2	explain workings of percentages		
3	show examples		
4	125; 0.1	850; 0.25	2950;0.75
5	any	any	any
6	show examples		
7	multiplication		

Lessons 4a–4c continued.

Q	4a	4b	4c
8	61 cm	113.8 cm	135.5 cm
9	children to show examples		
10	explanation of subtraction		

Lesson 5 (page 50)

A: 2 or 6; B: 80.5cm; C: class discussion

Lessons 5a–5c (pages 51–53)

Q	5a	5b	5c
1	nonogon	hendecagon	dodecagon
2	4	5	4
3	90°	60°	150°
4	5	20	40
5	cylinder, sphere, cone, hemisphere		
6	N/A	N/A	N/A
7	320	576	864
8	depends on how they are joined		
9	shapes depend on where they are cut		
10	children to extend pattern		

Lesson 6 (page 54)

A: class discussion; B: explanation showing division of sum; C: *n*th term = 5n

Lessons 6a–6c (pages 55–57)

Q	6a	6b	6c
1	show examples		
2	show workings of percentages		
3	any correct example		
4	e.g. 822,444,390	780,654,555	666,873,954
5	show workings of multiplication		
6	23*n*	32*n*	64*n*
7	show workings of addition		
8	e.g. 98 ÷ 5	198 ÷ 5	298 ÷ 5
9	any example that is correct		
10	e.g. A=8 B=15	A=9 B=25	A=126 B=223

Problems involving 'real life', money or measures

Lesson 1 (page 58)
A: 12996; B: 4.75kg; C: 25033m.

Lessons 1a–1c (pages 59–61)

Q	1a	1b	1c
1	£132.89	£267.11	£334.23
2	21	41	81
3	20;160	60;480	123;984
4	100 mins.	50 mins.	80 mins.
5	87	207	327
6	200 x 300	300 x 400	400 x 500
7	£41.71	£89.24	£206.74
8	£2860.25	£7110.25	£11,360.25
9			
10	40.84kg	33.84kg	59.84kg

Lesson 2 (page 62)
A: 1.7m^2; B: 0.9; C: 15:47.

Lessons 2a–2c (pages 62–65)

Q	2a	2b	2c
1	140m^2	600m^2	866.67m^2
2	10.5	14.7	20.5
3	£16	£26	£46
4	4 rows 42 not used	10 rows 12 not used	17 rows 27 not used
5	£25.50	£19.55	£36.55
6	Caitlin-2.4lbs	Caitlin-3.6lbs	Caitlin-4.8lbs
7	30cm	32.5cm	60cm
8	81.25	81.25	81.25
9	8	17	30
10	154.98m	217.98m	343.98m

Lesson 3 (page 66)
A: badminton and tennis;
B: 4 buses, 32 spare seats; C: £45,000.

Lessons 3a–3c (pages 67–69)

Q	3a	3b	3c
1	7/20	1/10	1/10
2	16	32	64
3	eg 4 x 6m	eg 4 x 9m	eg 6 x 9m
4	6.07kg	10.07kg	30.07kg
5	195	320	640
6	52.6cm	15.6cm	65.6cm
7	532mm	714mm	957mm
8	6:47 pm	7:29 pm	8:19 pm
9	none	5	13
10	2.04km	2.28km	2.52km

Lesson 4 (page 70)
A: £248.25; B: 114m^2; C: 67

Lessons 4a–4c (pages 71–73)

Q	4a	4b	4c
1	28 cups 4.2 litres	48 cups 7.2 litres	96 cups 14.4 litres
2	£41	£91	£141
3	960	3240	6480
4	£1 =1.47e	£1 = 1.51e	£1 =1.38e
5	2 minutes	2 min. 3 sec.	2 min.51sec
6	38.1cm	55.88cm	106.68cm
7	3kg	1.36kg	2.56kg
8	5.37 hrs	8.23hrs	13.94hrs
9	10.5	14.5	9.75
10	252 secs	1152 secs	2142 secs

Lesson 5 (page 74)

A: £182.63; B: 2816 litres; C: 252

Lessons 5a–5c (pages 75–77)

Q	5a	5b	5c
1	9.5hrs	11.5hrs	8.5hrs
2	£712	£1712	£2712
3	£43.70	£95.55	£320.55
4	3st 0lb	4st 0lb	6st 0lb
5	£85.25	£185.25	£385.25
6	2.75m	3.35m	1.38m
7	24th March 19:54	25th March 02:54	25th March 02:54.56
8	67.2 miles	201.6 miles	302.4 miles
9	30	60	80
10	£10.75	£14.50	£35.75

Lesson 6 (page 78)

A: 6 days, £92.50; B: 150g carrots, 125g potatoes, 475g chicken, 1.14 litres stock; C: £182, yes, skateboard costs £155.10

Lessons 6a–6c (pages 79–81)

Q	6a	6b	6c
1	walk 1/4 jog 1/2 run 1/4	walk 1/4 jog 5/8 run 1/8	walk 1/8 jog 3/4 run 1/8
2	140	85	349
3	socks £1.10 hat £2.50 gloves £1.60	£1.45 £2.25 £1.60	£4.50 £3.75 £1.60
4	any dimensions that give an area 124m²	192m²	268m²
5	27.5 hrs	147.5 hrs	219.5 hrs
6	12,200m	18,300m	36,600m
7	8	17	30
8	527	1079.5	1587.5
9	1845	3485	4305
10	£112.74	£190.29	£366.54

Lightning Source UK Ltd.
Milton Keynes UK
03 October 2009

144500UK00001B/8/A

9 781903 853795